After studying history at Edinburgh University, Alasdair Roberts worked extensively in Gaelic education, history and research. He is now a full-time writer in retirement from Aberdeen University.

Tim Kirby is a British artist and illustrator whose work has appeared in a large number of publications. His clients include *Gramophone* magazine, the *New Statesman*, *Management Today* and the BBC.

AND THEY WONDER
WHY WE DON'T COME
UP VERY OFTEN

Midges

Alasdair Roberts

ILLUSTRATED BY
Tim Kirby

BIRLINN

This edition first published in 2021 by
Birlinn Limited
West Newington House
10 Newington Road
EH9 1QS

www.birlinn.co.uk

ISBN: 978 1 78027 638 0

British Library Cataloguing-in-Publication Data
A catalogue record for this book is available from the British Library

Typeset by Mark Blackadder

Printed and bound by Bell & Bain Ltd, Glasgow

Contents

FLY-THRU

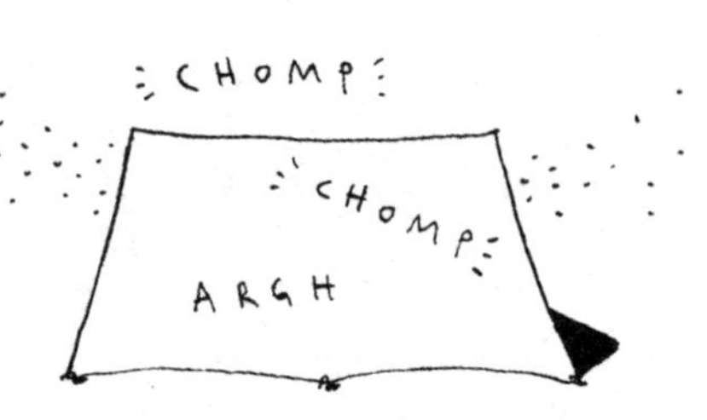

Personal

It is good to have a chance to update earlier editions of this little book. Lightly written and heavily illustrated, the book is still largely an anthology of midge quotes, lots of them funny on the banana-skin level of human suffering. The humour includes personal stories of midge attacks, but one thing which has not emerged clearly – till now – is my personal angle.

More than twenty years ago my wife and I camped our way slowly down to the south of France, prior to taking a house in the Languedoc and enrolling our children at the village school. I had been given leave from lecturing in order to write a book. At the simple campsites we preferred, there were sometimes rats and often mosquitoes. I attract biting insects, and the inside of the sleeping compartment became smeared with blood swatted from tiny predators. By the time we reached the Mediterranean I was swollen, itchy and groggy on antihistamine – but there, on the coastal strip, mosquitoes were no more. The French government had expelled them by draining the marshes where they bred, in order to create a string of holiday resorts between Marseille and Perpignan. My thoughts turned to what a Scottish government might do, and I wrote to Scotland's national newspaper about midges: 'Would this plague be tolerated on the south coast of England? Yet how many holidaymakers, well prepared for the West Highland weather, have vowed never to return?' Survey evidence, the first to be made public, suggests that the answer is 60 per cent.

Mostly I write history, so colleagues were surprised (some were scornful) when word of the latest project got out. 'Midges a modern scourge? The evidence of early accounts' appeared in a journal issued by the National Museums of Scotland. I knew that Johnson and Boswell had toured the Highlands in 1773, both recording the experience in print without saying a word about midges. I learned that a number of other outsiders, including scientists, had done the same and left a similar impression of a midge-free Scotland. In the 17th century there was also silence, the Gaels merely regarding *meanbh-chuileagan*, or tiny flies, as very small.

True, a military road-builder described midges

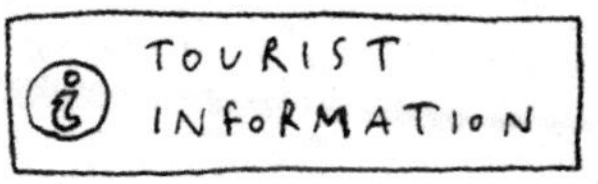

IS THAT ONE OF OUR VISITOR SURVEYS?

I THINK IT'S WRITTEN IN BLOOD

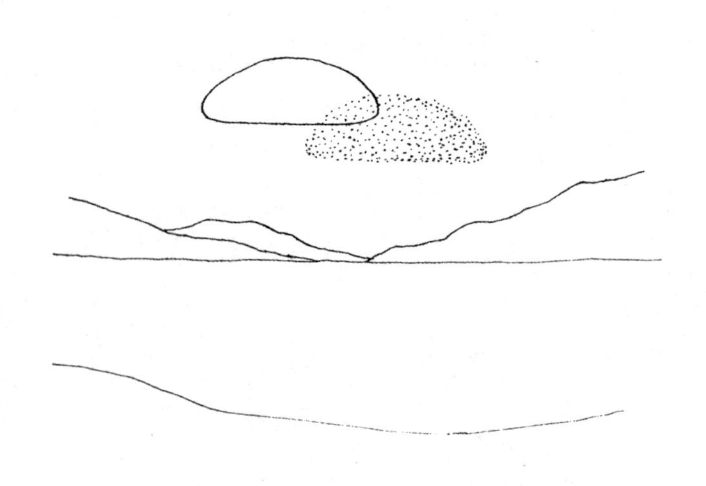

swarming before the Jacobite Rising of 1745; true also, Bonnie Prince Charlie was bitten during his summer in the heather after Culloden. But these were exceptions. And midges never featured during the next ninety years of military recruitment, emigration and visitors. Then everything changed in the middle of the 19th century: no account of grouse moors or salmon fishing in Victorian Scotland was complete without midges. I suggested a few environmental explanations linked to the Highland Clearances, inviting debate, and found myself addressing geographers in London. 'A rising cloud of midges in the Scottish Highlands?' attracted media attention, including a leader in *The Scotsman*: 'Just once in a while, a theory arises which can transform a society's whole perception of itself and of its forebears. There was Copernicus, Newton, Darwin, Einstein. Now there is Alasdair Roberts of Northern College in Aberdeen, and the thesis he has outlined to this week's Institute of British Geographers' conference.' Of course it was the third leader, which is often given over to humorous subjects. The writer went on: 'He accuses *Culicoides impunctatus* of being an incomer, a white settler of fairly recent provenance. In fact he suspects it of following the fashion for Highland holidays that took hold in the 19th century, and says it only gained its status in legend with the publication of a cartoon in *Punch* in 1853 which showed two artists being plagued by the familiar swarm.'

Midges have been in Scotland since the primeval bog, but never so many. Yes, I did raise the possibility that there was no real rise in numbers, just an increase in awareness, but that was not the conclusion reached. I really do believe that the midge problem is much worse than it was before Victoria and Albert took to the hills. Critics have queried the 'argument from silence'. One man at the London conference thought he'd squash me with an explorer who never mentioned mosquitoes – but had one flattened in his notebook. No one who has been targeted by a cloud of midges would compare the experience, for sheer must-mention drama, with the flea-bite level of other insects. The leader-writer's wit extended to the idea that Dr Johnson was saved by his well-known lack of personal hygiene, so that perhaps 'the midge is merely particular about what it eats'. Sweat attracts midges, but journalists will say anything in the silly season.

Christopher Smout, Scotland's Historiographer Royal, now concerns himself with environmental history. Contributors he brought together for a book about 'natural change and human impact' wrote chapters on the nibbling animal impact of both sheep

THE MIDGES ARE
VERY REALISTIC
WET
PAINT

and deer on the Highlands – and also climate change. Mark 3 midge-theory took in the cold 17th century as an additional factor. Scientists at Aberdeen University, where Alison Blackwell began her midge-researching career, were unimpressed by my proposal that the replacement of cattle-grazing crofters by sheep, and then deer, degraded the Highland environment. Having chosen trackless moors for investigating *C. impunctatus*, they were unlikely to give much thought to 'Midges in a changing Highland environment' (my chapter in the Smout book). Those Aberdeen scientists were also doubtful about climate change as a factor. Now global warming is accepted as a fact, explaining the modern rise in midge numbers – another new fact.

When the book came out we had just moved to an eight-acre croft on the west coast, where my wife reared curious breeds of sheep. Meanwhile I wrote Highland history and thought about the environment. Beyond our fence is nature stern and wild, but also a large area of lazy-beds or *feannagan* – abandoned cultivation rigs, raised by a foot or two so that rainwater could flow between them. Described as 'old arable' in the 1834 North Morar estate map, they still give a corrugated shape to the sloping ground. These lazy-beds (which were actually labour-intensive) testify to the fact that a large population was sustained here, mainly by potatoes, until blight and famine drove them out. As a historian I collect examples of large-scale drainage schemes by landowners like Cameron of Lochiel, but ordinary people cultivating every soggy acre are closer to my heart.

In those days the Rough Bounds from Knoydart to Moidart were almost entirely Catholic, and on an island in Loch Morar lived Bishop Hugh MacDonald, brother to the local chief. Island Morar was his 'cathedral', his base for travelling round the Catholic *Gaidhealtachd* from Badenoch to Benbecula and down through the Western Isles to Barra. Bishop Hugh built a fine house with a library of books in several languages, and opened a school for boys willing to follow his arduous profession. The last time I took a party out to the island the bracken was shoulder high and midges drove us back to the jetty. I played the pipes, emitting extra CO_2 (another attractor) while sweating. One lady, ankle deep in mud, lightened the mood with 'I'm a celebrity, get me out of here!' Island Morar could not have been like that in Bishop Hugh's day, which is the second personal reason (after the mosquito-free Mediterranean coast) for my conviction that something should be done about midges.

APPARENTLY ONLY THE FEMALES BITE
SOUNDS FAMILIAR
BUMPER BOOK OF MIDGES

Biology

'Midges. The scourge of the west, the terror of the Highlands. In itself "chust nothing at all"; *meanbh chuileag,* the tiny fly. Wingspan – 1.4 millimetres, but the numbers of them! By the hundred, and the thousand, lying and hatching, and up they rise to bite and suck and torment. One midge bite – nothing. Ten – a mere itch. But by the dozen, and the dozen every second, flying round your head, biting . . .

'Only the female midge bites. She bites not to eat, but as part of her reproductive cycle, and after she has begun pregnancy. Brave scientists assessed her biting rate in the post-war midge trials; hungry gravid lady midges attack in many dozens, and as many as 2000 to 3000 bites an hour were recorded'. [John Macleod, *The Herald,* 4 Sept 1993]

The Highland midge, *Culicoides impunctatus,* is found mainly in areas where annual rainfall exceeds 50 inches, so the west side of Scotland has the greatest reputation for them. In 1990 220 inches fell on Lochaber. Thirty days is long life to a midge but new generations appear through summer. The first-born are males which do not bite, and it is usually June before females go in search of blood-meals. Male midges feed on plant juices. The biting season is only about ten weeks but so is the tourist season.

Midge larvae spend ten months in a non-feeding state from one year to the next, but then develop from the state of tiny worms to tiny adults in a few days. Up to eighty eggs are produced in a first batch, using energy stored from the non-feeding stage. Mating takes place when the male's antennae detect the high-pitched vibration of the female's wings, triumphant after a blood-meal. A second batch of eggs can then be laid. Females find blood by detecting the carbon dioxide breathed out by humans, horses, cattle and deer. Midge 'bites' are produced when the insect's mouth-parts use a scissor-like movement to rasp, not pierce, the outer skin. Hollow like a hypodermic needle, these then penetrate the tissues under the skin.

The scientist Alison Blackwell assured the author that women are more at risk than men. Not in the author's family they aren't. But people vary in their attractiveness to midges, and partly after being bitten already. No, this is not like inoculation but the reverse. Visitors have an immune system which does not respond to midge bites for a few days. Locals who risk their skin year in year out bringing in the hay are just as likely to be affected as tourists: 'If it is any consolation there are a great many Highlanders who, to their dying day, suffer just as much as the tourist after the first week of a holiday in the area.' [George Hendry, *Midges in Scotland*, 1987]

MUCHUS
BITUS

History

Martin Martin of Skye, a late 17th-century preacher, had a keen eye for flora and fauna including the least of God's creatures: 'The Land, and the Sea that encompasses it, produce many things useful and curious in their kind, several of which have not hitherto been mention'd by the Learned. This may afford the Theorist Subject of Contemplation, since every Plant of the Field, every Fiber of each Plant, and the least Particle of the smallest Insect, carries with it the impress of its Maker.' Midges are never mentioned in his *Description of the Western Islands of Scotland*.

The keeper of Oxford's Ashmolean Museum Edward Lhuyd reached Scotland in 1699 on his tour of Celtic countries. Lhuyd's collection of 'charms for preventing diseases in man or beast' recorded nothing against midges, and the Gaelic name for the midge, *meanbh chuileag* or tiny fly, has no threatening connotation. The bards are silent. The Blind Harper never sang of midges at Dunvegan. Alasdair Mac Mhaighstir Alasdair encountered none between Knoydart and Ardnamurchan. In his Song of Summer the only insects are *beachach* [bees] and *seilleanach* [wasps] despite perfect conditions for midges – *ciurach, dealtach, trom, blath* [misty, dewy, heavy, warm]. Even when the poet calls down biting insects on Eigneig and its anti-Jacobite priest, there are no midges: 'And every beast that on me preyed – The wasp, the gadfly and the bee.' The *creathlag* or gadfly is that other Highland pest, the cleg. Timothy Pont's 16th-century map of Edrachillis proclaimed, 'All heir ar black flies in this wood . . . seene souking me[n]s blood' – again surely clegs.

Edward Burt, a Londoner, built military roads to control the Highlands after the Jacobite Rising of 1715. Here he is near Fort Augustus showing that biting midges could be encountered if the landscape was sufficiently disturbed:

I have but one Thing more to take Notice of

in relation to the Spot of which I have been so long speaking, and that is, I have been sometimes vexed with a little Plague (if I may use the Expression), but do not you think I am too grave upon the Subject; there are great Swarms of little Flies which the Natives call *Malhoulakins*: *Houlak*, they tell me, signifies in the Country Language, a *Fly*, and *Houlakin* is the Diminutive of that Name. These are so very small, that, separately, they are but just perceptible and that is all; and, being of a blackish colour, when a number of them settle on the skin, they make it look as if it was dirty; there they bore with their little augers into the pores, and change the face from black to red.

They are only troublesome (I should say intolerable) in Summer, when there is a profound Calm; for the least Breath of Wind immediately disperses them; and the only Refuge from them is the House, into which I never knew them to enter. Sometimes, when I have been talking to any one, I have (though with the utmost Self-denial) endured their Stings to watch his Face, and see how long they would suffer him to be quiet; but in

> three or four Seconds, he has slapped his Hand upon his Face, and in great Wrath cursed the little Vermin.
> [*Letters from a Gentleman in the North of Scotland*, 1754]

Outsiders became fascinated in the 18th century by *Ossian,* James MacPherson's tales of a legendary Gaelic past. Literary people came north, some already devoted to high lands like the Wordsworths. Enthusiasm for the Highland Tour was a European phenomenon. Visitors kept diaries, many found publishers, but none mentioned midge bites. Thomas Thornton, an early sportsman in the Highlands, pursued a great variety of birds, beasts and fish during the 1780s: 'July 26. Day charming. Went to some lochs . . . Said to be six miles off, but turned out ten. The day was too calm.' Twenty-seven trout were caught to provide a loch-side meal at the end of the day. No biting insects disturbed the sportsman or his ghillies while they ate. Compare this with a modern fisherman's tale.

We pitched camp in the dusk on the *bealach* [pass] between Loch Hourn and Loch Quoich beside the tiny *Loch a' Choire Bheithe*. Although the normally boggy hollow was merely moist after the drought, it was still an idiotic choice, harbouring the most feared of Highland predators. As we pitched our tents clouds of midges joyfully attacked the source of fresh blood, making our task a misery. At least I had the foresight to keep my inner tent zipped tight. John tried, very stoically, to fish the lochan but merely served as dessert for the thickening swarms of tiny winged beasts.

Meanwhile I had built a fire of bogwood on the largest and flattest boulder and sought sanctuary from the midges in the thick drifting smoke. Once John had abandoned his fishing we supped and talked and drank on our smokey boulder and forgot the midges. They had remembered us. As I doused the fire and packed away the food, John tried washing in the burn. The midges regrouped and surrounded him as he tried to get dry, dressed and into his tent. He hardly tarried, but those few moments saw him bitten scarlet and his tent filled with midges. During the rest of

> the night the threshing of body and limbs
> and the dejected groans from his tent
> became increasingly pitiful.
> [James McLeod, *The Scotsman*, 29 June 1985]

James Hogg, the Ettrick Shepherd, experienced bed-bugs in an inn, which shows that journal-writers did not ignore insects:

> I got the best bed, but it was extremely hard,
> and the clothes had not the smell of roses.
> It was also inhabited by a number of little
> insects common enough in such places,
> and no sooner had I made a lodgement in
> their hereditary domains than I was attacked
> by a thousand strong.
> [*A Journey through the Highlands,* 1804]

Robert Burns, breaking his Highland journey on the notoriously midgy banks of Loch Lomond, found nothing to complain of. He spent a night of festivity in the mansion of a Highland gentleman and used the swarming of tiny creatures to convey a cheerful image of Scottish country dancing: 'The ladies sung Scotch songs like angels, at intervals; then we flew at Bab the Bowster, Tullochgorum, Loch Errol Side, &c, like midges sporting in the mottie sun . . .' [*Tours of the Highlands and Stirlingshire*, 1787]

The midges which Burns knew, dancing in dusty sunlight, were obviously the garden variety [*C. obsoletus*] which scarcely bite. In Lowland Scotland as well as England 'midge' has always meant tiny, with no suggestion of biting. Scripture supports this. In Matt. xxiii, 24 fastidious Pharisees are condemned as 'Blind guides! You strain off a midge and gulp down a camel!' or in the Scots Bible 'clengeand a myge bot suelliand a camele.' The same relaxed attitude was common in the Highlands where, as a proverb had it, 'The cow is only a good deal bigger than the midge.'

The argument that Highland midges were little known in the early 19th century gains support from the silence of Professor John Wilson of Edinburgh University. In an account of a walking tour in the 1820s he wrote 127 pages on 'The Moors' without mentioning them. He did mention ants:

Go to a desert and clap your back against a

> cliff. Do you think yourself alone? What a ninny! Your great clumsy splay feet are bruising to death a batch of beetles. See that spider whom you have widowed, running up and down your elegant leg. Meanwhile your shoulders have crushed a colony of small red ants settled in a moss city beautifully roofed with lichens – and that accounts for the sharp tickling behind your ear, which you keep scratching . . . All the while you are supposing yourself alone! But the whole wilderness – as you choose to call it – is crawling with various life. [*Recreations of Christopher North*, 1865]

Thirty years later Charles Weld provided striking contrast on a Caithness fishing trip:

> Talk of solitude on the moors! – why, every square yard contains a population of millions of these little harpies, that pump blood out of you with amazing savageness and insatiability. Where they come from is a puzzle. While you are in motion not one is visible, but when you stop a mist seems to curl about your feet

> and legs, rising, and at the same time expanding, until you become painfully sensible that the appearance is due to a cloud of gnats. When seven miles from Scourie I came to the Laxford, a glorious salmon river spanned by a bridge, backed by Ben Stack and framed by rocks, garlanded by fern and birch. A lovely subject for a sketch, but, in my case, unsketchable, for I had no longer sat down than up rose millions of midges, which sent me reeling down the craggy steep, half mad. [*Two Months in the Highlands*, 1860]

When army engineers turned from road-building to map-making the enemy was there in squadrons:

> The heat also then being intense above Loch Maree and Gairloch, it was our practice in walking to put our coats and waistcoats into our knapsacks, and thus, with our shirt necks thrown open, and our sleeves tucked up, we were exposed in a peculiar manner to the baneful attacks of those venomous insects. On the occasion referred to we suffered very

severely; our arms, necks and faces were covered with scarlet pimples, and we lost several hours' rest at night from the intense itching and pain which they caused. Even at the inns we had frequently to smoke in our bedrooms and over our meals to drive these insects away.
[J. E. Portlock, *Major-General Colby*, 1869]

NONE
SHALL
PASS

Environment

The father of the conservation movement in Scotland, Frank Fraser Darling, saw a problem without a solution:

> Almost everywhere in the Highlands below 2,000 feet there are vast hordes of midges which affect the movements of mammalian life, including man, to a considerable extent. The place of the midge in human ecology is such that a greatly increased tourist industry to the West Highlands could be encouraged if the midge could be controlled. But every square yard of Highland and Island moors has its midges. Little if anything can be done by way of control which would not cause extensive damage to agriculture, forestry, game and freshwater fisheries.'
> [*The Highlands of Scotland*, 1964]

The French government succeeded in eliminating the mosquito west of Marseille in the 1970s, turning their south-west Mediterranean coastline into another Côte d'Azur. Scotland's tourist industry proclaims that 'One visit is never enough' – but it may be. HOST, the former Highlands of Scotland Tourist Board, did no marketing research on whether midges deter return visits, despite dark rumours of suppressed evidence. While monitoring everything from bed nights to bus parties, the question was never put. Perhaps it only applies to those visitors – their accommodation ranging from youth hostels to shooting lodges – who come for the outdoor environment. Alison Blackwell's research among campers and caravaners found 86 per cent of first-time visitors would not recommend Scotland in summer, and £286 million a year is regularly quoted nowadays as the cost to Scottish tourism. It seems to be based on all visitors staying an extra day.

When a brochure was produced featuring a midge blown up 500 times as a 'welcome pack' for hotels and guest houses, reaction in the tourist industry was understandably hostile. Landladies didn't like it, and the manager of a visitor centre in the wettest, midgiest

NO, WE DON'T GET THOSE HERE. IT'S, ER, PEPPER

corner of Lochaber refused to stock the first edition of this book because 'We're not in the business of encouraging the midge myth here.' At national level the Visit Scotland response has been on the lines of 'there have been no recorded deaths due to a midge bite so far as I know' – as if anyone ever suggested that single midge bites were the problem. Civil war finally broke out in summer 2004 when Grampian, Tayside and Fife Tourist Boards told visitors that 'if you want to avoid being bitten alive on holiday then head east instead of west'. Again a giant midge featured, this time with someone inside the costume.

When a stretch of new road was built on the notorious A830 between Arisaig and Morar, environmental experts were on hand to ensure that the boggy Moss of Keppoch stayed that way, although a 19th-century landowner made strenuous efforts to drain it. Conservationists resist the drainage of breeding grounds and see the midge as 'a guardian of wildness in the Highlands' linked – somehow – with 'restoring the Caledonian Forest'. [*Caldedonia Wild!* Winter 2002–03] They are probably unaware of the extent to which the 'unspoiled' wilderness has been created by man. Loss of trees led to an increase of water content in the soil and more acidic breeding grounds. The Highland Clearances, removing people in favour of sheep in the first half of the 19th century, seem to have been associated with an increase in the midge population. Large tracts were subjected to heather burning for the sake of increased sheep pasture. James Hunter has argued that the old agricultural system bequeathed to the new sheep graziers upland pastures fertilised by cattle. The 'sheep-sick' environment was discussed when lambing percentages fell later, but it was in the 1850s that deterioration of land began due to 'purely extractive' sheep farming. The grazier Charles MacLean emigrated from Morar to Gippsland in the Australian state of Victoria, replacing the aboriginal population with merino sheep. His brother John MacLean used up the best of Glenmoidart's pasture for his flocks before doing the same on the isles of Canna and Eigg.

'It is natural, in traversing this gloom of desolation, to inquire whether . . . those hills and moors that afford heath cannot with a little care and labour bear something better? The first thought that occurs is to cover them with trees.' Thus Samuel Johnson – but what kind of trees? High ground was seeded with conifers last century but there is a difference between natural woodland and forestry plantation. The loss of natural woodland and the rise of forestry estates have combined to provide habitats congenial to midges, with a reduction in the dragonflies and frogs which prey on larvae. Forestry gives shelter from wind and shade from sunlight. Bracken also favours the insects. It spreads where arable land is neglected and offers shade.

We will never recover 'the great forest of Caledon', but woodland can return. Sir John Lister-Kaye suggests that estate owners should give up 15 per cent of their hill units to natural restoration for 25 years. Aware that sheep and deer have 'enormously depleted the natural fertility of the original woodlands', Sir John visualises 'a jigsaw of restoration right across the Highlands and Islands'. Prince Charles agrees: 'The type of sustainable land management ethic advocated by John Lister-Kaye is, I believe, a key part of a more balanced and long term approach to the management of the fragile Highland ecosystem and the equally fragile human communities which depend upon it.'

Climate change came to the Atlantic edge of Europe in the 17th century. A lowering of sea temperatures brought down Scotland's snow-line by 400 metres, and that greatly reduced the area available for breeding. Cold winters affect the survival of midge larvae and chilly summers the eggs born of blood meals. Long before global warming, temperatures rose in the early 18th century – hence Burt's 'troublesome' midges.

IT'S NOT THE ENEMY I'M BOTHERED ABOUT, IT'S THE MIDGES

Prince in the Heather

Charles Edward Stuart made an enforced and extended tour of the Highlands and Islands after the defeat at Culloden in 1746, and was bitten by *Culicoides impunctatus*. In June the no longer bonny Prince Charlie was in Uist, sheltering under a sail in the heather near Loch Boisdale. One of his party wrote of the ordeal, 'We were never a day or night without rain; the Prince was in a terrible condition, his legs & thy's cut all over from the bryers; the mitches or flys wch are terrible in yt contry, devored him, & made him scratch those scars, wch made him appear as if he was cover'd with ulsers.' [A. & H. Tayler, *1745 and After*, 1938] The playwright George Rosie made a link with the Prince's broken-down condition in 1784, when Charles held court in Florence: 'His formerly superb physique is in a state of collapse. He has scurvy sores on his legs which cannot be healed, and cause him constant pain.' [*Carluccio and the Queen of Hearts*, 1992]

'Over the Sea to Skye' seemed the right idea, but midges were not finished with the royal fugitive. John MacDonald of Borrodale was with his Prince in Glenmoriston on 28 July:

> The evening being very calm and warm, we greatly suffered by mitches, a species of little creatures troublesome and numerous in the highlands; to preserve him from such troublesome guests, we wrapt him head and feet in his plead, and covered him with long heather that naturally grew about a bit of hollow ground where we laid him. After leaving him in that posture, he uttered several heavy sighes and groands. We planted ourselves about the best we could.
> [Bishop Forbes, *The Lyon in Mourning*, 1896]

The Catholic prince was not keen on devotional books, or he might have thought of praying to St Jacobus. This saint brought down clouds of gnats on an army besieging the city of Nisibis and put it to flight. The Jacobites could have used his help against Hanoverian artillery.

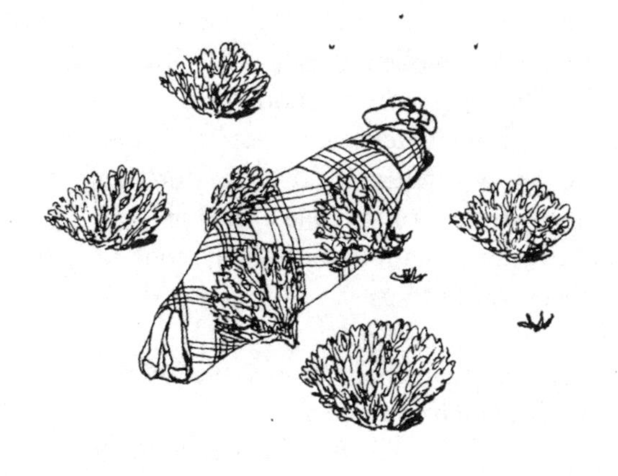

Tearlach took to smoking a pipe. Tobacco was a luxury beyond most Gaels, but their lives were permeated by smoke, as visitors noted. John Leyden: 'The huts of the peasants on Mull are most deplorable . . . There is often no other outlet of smoke but at the door, the consequence of which is that the women are more squalid and dirty than the men, and their features more disagreeable.' [*Tour in the Highlands,* 1800] Robert Southey: 'The smoke is clean, and the smell, to me at least, rather agreeable than otherwise: but it attacks the eyes immediately, and that it injures them is plainly shown by the blear eyes which are here so common among old people.' [*Tour in Scotland,* 1819] J. E. Bowman, a banker on holiday, took his readers inside a black house:

> It was formed wholly of turf, the walls not five feet high, and the roof very steep, particularly at one end, where it rose a little higher in a conical shape and ended in a smoke hole . . . It is a singular though well ascertained fact that bugs are never found, even in the greatest filth, where peat is used. [*The Highlands and Islands,* 1825]

Most visitors regarded the natives as dirty, but in their clothes, hair and skin Gaels were smoked like the herrings they took from sea lochs. Perhaps Highlanders made brown by turf fires were protected. Ronald Black, a Gaelic expert, thinks so: 'My impression is that midges didn't greatly bother the old Highlanders: while not unwashed, their skin was certainly kippered by peat-reek, as old photographs show.'

COME ALONG HAMISH, KEEP UP

Kilts

Ragnall Mac Ille Duibh (to give Ronald Black's Gaelic name) has another explanation for the Highlander's tolerance of midges: 'The traditional plaid was both clothing and bedding, and could be drawn about every part of the body as required. That's why the proscription of Highland dress in 1747 aroused more fury than any legislation before or since.' Earlier the ironmaster Thomas Rawlinson persuaded his workers to wear the small kilt, or philabeg, while working in the foundry at Invergarry. Cynics believe that this Englishman invented the kilt. It eventually replaced the enveloping philamore when Sir Walter Scott reinvented tartan – through his books, and on the ample frame of George IV on a royal visit to Edinburgh. The monarch's best-dressed friend in youth had been Beau Brummel. In middle age he took the precaution of wearing flesh-coloured tights.

MIDGES

Para Handy, skipper of the puffer *Vital Spark*, has a kilt story concerning 'mudges':

> What wass there on the island at the time but a chenuine English towerist, wi' a capital red kilt, and, man! but he wass green! He wass that green, the coos of Colonsay would go mooin' along the road efter him, thinkin' he wass gress . . . The first night on the island he went out in his kilt, and came back in half an oor to the inns wi' his legs fair peetiful! There iss nothing that the midges like to see among them better than an English towerist with a kilt: the very top wass eaten off his stockins'.
> [Neil Munro, *Tales of Para Handy*, 1931]

It is not true that the redcoat soldiers at Culloden bayoneted kilted opponents who needed a hand free for slapping bare legs. Nor is it accurate to say that midges caused the Highland Fling. But there is no doubt that English workmen were brought to the island of Rum to build a castle for the Lancashire industrialist Sir George Bullough and were required, despite their Accrington origins, to wear his Rum tartan kilt. The midges must have been delighted by this prime English leg. Sir George only prevented a strike by agreeing to pay the men a tobacco allowance of tuppence a week.

Kinloch Castle is now well known through television as an Edwardian pleasure palace. Archie Cameron who grew up on the island knew more about pain: 'Apart from the factors, the chief bane of life on Rhum was certainly the midges, which seemed to be a particularly virulent and persistent strain. Experts say that midges don't like sunshine and don't go out to sea, and the experts are wrong . . . There was one day I was passing a field where three men were scything hay. It was a very hot day, with brilliant sunshine, and I knew the men must be very uncomfortable for they were wearing fine muslin veils as protection against the midges . . . What puzzled me that particular day, though, was that each man seemed to be carrying a sheet of glass on his back. When I move closer to satisfy my curiosity on this I was amazed to find that it was the sun shining on the wings of a solid mass of midges . . .

'There was one case where a man died from

their bites. It was long before my time and indeed happened in the old days, but the story was still repeated and there was some evidence that it was true. There is an old road to Kilmory from Kinloch, on the north side of the glen, and half way along there stands a small uninscribed headstone which we knew as "The Baby's Grave". This was one of our favourite walks, and often enough I heard my father tell the story as he stood there.

'It was all long ago, in the time of the chiefs, and when the baby daughter of the chief died he instructed two of his trusted men to carry the coffin to Kilmory for burial. [It's true that stillborn unbaptised babies were buried without ceremony in designated places.] With only half their journey completed a severe thunderstorm came on, and the two men decided to go no further but to bury the baby where they were and say nothing about it. But the two men fell out about something and the facts were revealed.

'The chief was exceedingly angry, especially at the man who had been in charge. He was stripped and pegged out on the grass at the place where the school now stands. The chief's wife, still mourning the death of her baby daughter, believed that such revenge was too harsh and pleaded with her husband for mercy. He relented and the man was released, but too late, for he had been so sorely bitten by the midges of Rhum that he died in agony anyway.'

WE ARE NOT AMUSED

Royalty

The Sutherland estates were the first to experience clearance of the native population, along with their traditional farming. Queen Victoria paid a visit to Dunrobin Castle in 1872, when vast numbers of sheep and deer had been nibbling for more than half a century: 'The heather is very rich all round here. We got out and went into it . . . drove down again, and before we were out of the lower wood, which is close down upon the sea-shore, we stopped to take our tea and coffee but were half devoured by midges.' [Queen Victoria, *Leaves from Our Life in the Highlands,* 1868] Balmoral was chosen as a royal residence because it was in the drier half of the Highlands, but it also had midges:

> 'She'll insist on taking you to her "little bothy"
> at Glassalt Shiel. Make sure you cover up,
> because the midges there are without

> exception the most voracious that I have ever come across.' He was right. The only known midge repellants were paraffin oil and tobacco smoke, and Rachel was more than happy to follow her sovereign's stout, black-bonneted example and puff and choke away at one cigarette after another.
> [Reay Tannahill, *In Still and Stormy Waters*, 1992]

The novelist exaggerates. Victoria discouraged smoking at Balmoral, but once, when John Brown was not looking, she borrowed a great-grand-daughter's cigarette in an attempt to keep midges at bay.

The *Aberdeen Press and Journal* takes special interest in the Royal Family when they return to Balmoral each summer:

> The infamous Scottish midgie is no respector of persons – as the Princess Royal found to her cost yesterday. Every swarm of midgies from miles around seemed keen to join the crowds as the princess started her helicopter-

> hopping tour of the Highlands at Kyle of Lochalsh. Soon business-men, councillors and officials were literally itching to be presented to her royal highness. And the temptation to swipe away the insects soon overcame etiquette. But blue blood is no protection against the attentions of the pests. The princess, too, had to resort to the occasional rubbing of the royal cheeks as she spoke with her guests.

This reference to Princess Anne's face recalls an assault on her granny. A minister's wife came to the rescue when the Queen Mother was attacked during Highland Games at the Castle of Mey, her country retreat in Caithness. People noticed the Queen Mother was being bitten when she began waving the tiny insects away from her face. A woman nearby whipped out a midge spray from her handbag and applied it, for some reason, to the royal legs . . . Later Anne made her views public at the Loch Lomond Visitor Centre: 'Visitors to Scotland see the country as near perfect, apart from the X-factor of course – the midge.'

Nature

Nowadays nobody believes that midges should be attacked with chemical weapons. We have all become aware of how pesticides enter the food chain, threatening various species including ourselves. Do midges do any good, environmentally speaking? On the Russian steppes midges and other blackflies may save the fragile tundra by keeping grazing animals on the move. In Scotland, the grazing patterns of red deer are partly controlled by the midges: 'In summer the severe attacks force the deer up to the higher and more windswept hills.' [Hendry, *Midges in Scotland*] Perhaps more birds in the Highlands would help. When a record number of pied fly-catchers was reported in the Inversnaid reserve at Loch Lomond an RSPB spokesman told the *Glasgow Evening Times*: 'With the swarms of midges around at the moment, which flycatchers eat, the more "pied flies" the better!'

The news of a parasitic red mite, small enough for four to live off a single midge and introduced illicitly from Africa by entomologist Eugen Hayek, sounded like a spoof in *Private Eye*. In November 1992 Lord Fraser of Carmyllie, Minister of State for the Scottish Office, told the Upper House: 'The Red Mite is proving a mite elusive . . . We are not actually sure it exists.' In the following May, however, Britain's traditional landowners were told by a 'red-faced' Lord Fraser that research into its habits was being conducted on behalf of the Natural History Museum. Lord Campbell of Croy saw this as a victory. 'I'm delighted there are people concerned about these beasties,' he told a mirthful House to cries of 'What about the Blandford Fly?'

Anne Baker was funded by the British Entomological Society and spent the summer of 1993 in Skye collecting midges. Only 2 per cent of her sample were weakened by red mites, and 'experts believed that it was not in the interest of the parasite mite to weaken the host midge to the point of death'. This seems a mite obvious to the non-expert. Healthy midges achieve a grand old age of thirty days, with those carrying the dreaded red mite cut down in their

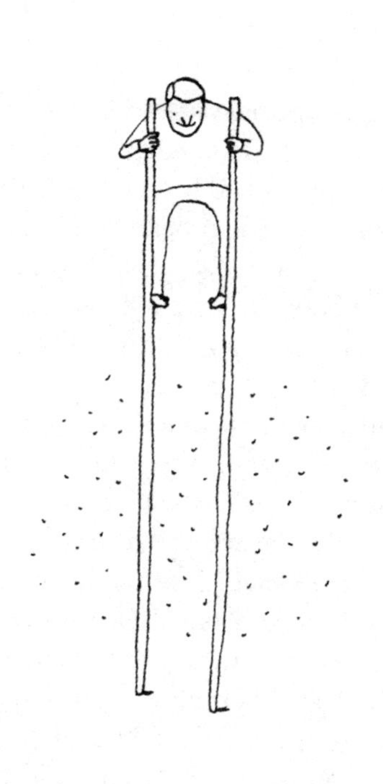

twenties. Chronic illness in one midge out of fifty seems a very small step for man.

Science came together with archaeology when it was reported in 1993 that midges and their much-publicised parasites had been in the Highlands since the dawn of time. The programme *Tomorrow's World* announced that a midge fossil had been found in resin, with a mite still attached to it, from 70 million years ago. Lord Fraser saw this as bad news: 'If over that extended period the mite has been unable to bring to an end the primitive midge, it is unlikely within your Lordships' lifetime to achieve any success against the ferocious Highland midge.' But the fossil offers hope that the analysis techniques of archaeology, which have yielded information on insects larger than midges, may come up with evidence about change over the centuries. The answer lies in the peat bog, but Dr Nick Dickson of Edinburgh University denied that he had found midges preserved in a prehistoric crannog or lake-dwelling in Loch Tay.

Linked with red mites, at least in the House of Lords, are bats. Lord Fraser admitted that the Natterer's Bat sounded 'something like a refugee from a mother-in-law joke', but was able to reveal that it did exist, favouring larger insects while accepting midges too. The Pipistrelle bat, however, is said to be able to eat a thousand midges a night, or twice its own weight. Bat-houses among the trees at Nethybridge golf course have transformed a notoriously midgy stretch of the fairway. In the absence of belfries, bats need more dwelling houses and byres than the Highlands have seen since the Clearances. Is this part of the reason, along with peat smoke, why Gaels were not bothered by midges near home?

Science

After the last war when government was considering what might be done, Dr Douglas Kettle of Glasgow University studied breeding habits near Luss on the banks of Loch Lomond. Other samples were gathered from a 'moorland site' near Glasgow – Drumchapel. Kettle's main interest was in larvae which could be studied out of the biting season. Most were found near the surface of wet soil in bogs, and it was easy enough to bring back samples to the lab and separate out worms with sieves and concentrated magnesium sulphate. '*Culicoides* larvae were readily recognised by their active lashing swimming movements.'

Today animal rights activists would be out with placards: 'Due to the exhaustion of the food supply the larvae failed to grow, although they often survived for long periods.' The phrase 'larval preservation' meant being killed by the aptly named Dr Kettle: 'Living larvae, dropped into hot water were preserved in 70 per cent alcohol.' Later Kettle had very limited success in his attempts to eradicate larvae and midges on their home ground by chemical means. [*Bulletin of Entomological Research*, 1957] Modern researchers have found that pure concentrate of garlic oil, with Allicin the active ingredient, is only 40 per cent effective against larvae in nature but entirely so in the lab – where it acts like nerve poison and creates the smell of an Italian restaurant.

In the 1990s scientists began going on summer safari led by Dr Alison Blackwell of, in turn, the universities of Aberdeen, Dundee and Edinburgh. Even in winter Alison's arms showed the scars, until a protective outfit was devised to make field-workers look like beekeepers. Bugwear, which sells at £65, has panels of fine mesh, extra long sleeves that can be worn over the hands and a stand-up collar with hood attached. A baseball cap is recommended to keep the hood away from the face. This is clearly an advance on muslin: 'Soon our veils were almost airtight . . . It was either suffocate or breathe midges and scratch mud and blood. We lads wore knickerbockers, and the gap between them and the top of our stockings was a favourite feeding ground.' [Archie Cameron, *Bare Feet and Tackety Boots*]

> Scientists working in Argyllshire believe they have discovered the perfect midge bait, a chemical call sign emitted by the insects which alerts others to the prospect of a good meal . . . The researchers believe a synthetic version of the chemical, known as a recruiting pheromone, could be used to lure *Culicoides Impunctatus Goetghbuer* into traps by the million.
> [*Sunday Times*, 13 June 1993]

There were two study sites: Ormsary, Argyllshire, and Kinlochewe in Wester Ross. Searches for swarms were made around dawn and during the three to four hours before dusk. About a third of the swarms were netted completely and numbers counted. Precise data emerged for the first time. The electron microscope was used on freeze-dried midges, painted gold. Females turn out to have three times as many olfactory sensilla as tactile hairs on their antennae, reflecting their dependence on smell for finding blood meals. Swarms ranged from single figures to an uncountable one of 3–4,000 midges. The commonest shape was oval and up to two metres wide. Swarms were videoed to examine flight paths, traced at 40 millisecond intervals by a colour monitor:

> Each midge flew with a fast zigzagging movement, falling to the bottom of the swarm, circling back up again, pausing briefly and falling again – all in about one second. Midges faced upwind. Downwind drift was recorded, and in gusts of wind swarms would be blown away. Light levels were the primary stimulus for swarm formation. Most swarms occurred one or two hours before sunset in low light, with swarm numbers increasing in still, humid conditions. In ideal climatic conditions *C. impunctatus* swarms were observed for almost an hour.
> [*Ecological Entomology*, 1992]

The tendency of midges to emerge before dusk in conditions of low wind was confirmed by observing insects normally too small for the tracking and trapping techniques of entomology.

Another scientist admitted ignorance: 'The fine details of the mating process are not known, because of the inherent problem of trying to monitor the goings-on of a couple of busy insects only two millimetres long.' [Hendry, *Midges in Scotland*] After Alison Blackwell the secrets are out:

> For males to wait in the vegetation for emerging females may, at times, be sufficient to effect mating. Males would then expend less energy than for swarming and would be sure to mate with young, virgin females which would be maximally receptive. Swarming helps mating. Pairs presumably form within the swarms and complete mating on the ground or a nearby surface. A total of thirty-six copulating *C. impunctatus* pairs were observed on warm still evenings. Pairs were never seen actually forming, and therefore the copulation times recorded are shorter than the mating times. Some pairs separated almost immediately. Most copulation periods were one to three minutes, although one pair remained united for eighteen minutes. [*Ecological Entomology*, 1992]

Evidence has emerged from Edinburgh University's Centre for Tropical Veterinary Medicine that some people are more prone to midge attacks than others . . . 'This depends on the behavioural and electro-physiological responses of midges to different people's sweat extracts . . . The discoveries about how midges' antennae respond to human sweat could ultimately help entomologists develop new repellents to block receptor sites.' (*EditEd*, Spring 2002)

Early in the modern phase of midge research it became clear that what had previously been seen as two species were in fact two generations in the same summer. Following the warm August of 2002, however, the Blackwell research team was announcing a third generation liable to appear every three or four years. Global warming began to be accepted by scientists around the turn of the millennium, along with the fact that rising midge populations are a result.

WEE
BITE
FURIOUS
GNASHINGS

Places

'Is it the midges of Skye that are most ferocious? Or the beasties of Achnashellach? Are they more numerous in Glenelg, the bites more painful by Strontian? Midges seem less active in the Outer Isles, where still nights are infrequent – but when they come they bite – and how.' Thus John MacLeod from the Outer Isles, raising a question for anyone planning a holiday route.

Hamish Henderson spent much of his life roaming the country in search of tinkers – travelling people – or anyone who remembered an old song or story for the School of Scottish Studies in Edinburgh. His own midge memories went back to childhood:

> Roaming around the Perthshire Highlands on a push-bike when I was a kid, it always seemed to me that these areas were less midge-infested than areas north and west of the Great Glen. I remember pitching a bivvy at Strome Ferry on

> a spot which turned out to be their GHQ; some of them got inside the bivvy, and in ten minutes I was nearly bitten to death. There was nothing for it but to abandon the position.
> This really is an important question from the 'courtesy to tourists' point of view. Visitors to Scotland are entitled to information about the real danger-zones so that they can come prepared – or at any rate forewarned.

He is right: visitors could and should be given advice about where to avoid. We are familiar with ski reports: why not midge reports?, I asked in the first edition. They are there now: 'In the summer of 1996 the newly launched Lochbroom FM was in the headlines when, alongside weather reports and shipping news, it offered an early morning midge count. On a scale of one to ten a local seer would be called upon to assess the dangers of walking abroad at dusk.' [*Sunday Times*, 7 June 1998] Nowadays an inter-net visit to www.midge forecast.co.uk does the trick.

Argyll accounts for a large part of the West Highlands. Making the best of it, local artists with a 'midge-sized budget and a midge-sized office tucked behind Inverary Jail' hold a six-week midge festival. The Ugly Bug Ball and Midge Summer Night Ceilidh are highlights, with

UGLY BUG BALL

SO, DO YOU COME HERE OFTEN?

Mardi Gras-style processions through places like Campbeltown and Lochgilphead. The West Highland Way from Loch Lomond to Fort William attracts more walkers every year, and the main problem is treated on the website with similar 'Look on the bite side' humour: 'Anyway you'll be doing your bite – I mean "bit" – for ecology because if you walk the West Highland Way you will help the next generation of midges get a good start in life. Awesome!' Knowledge of the 'West Coast's most notorious midge havens' produced a litany for the *Scotsman* Diary's midge anecdotes competition about 'the flesh pots of Arrochar and Inverary, the pure-bred specimens of Crarae and Minard, the seasonal gatherings at Lochgilphead and Ardrishaig and the ferocity of Tarbert's denizens, particularly during a wet Fair Week. Thereafter, the choice is between the ear-piercing fraternity of Tayinloan or the orifice-invaders of Carradale.' The crew of the *Vital Spark* had their own memorable landings to conjure with:

> 'The Congo's no to be compared wi' the West of Scotland when ye come to insects,' said Para Handy. 'There's places here that's chust deplorable whenever the weather's the least bit warm. Look at Tighnabruaich! – they're

that bad there, they'll bite their way through corrugated iron roofs to get at ye! Take Clynder, again, or any other place in the Gareloch, and ye'll see the old ones leadin' roon the young ones, learnin' them the proper grips. There iss a special kind of mudge in Dervaig, in the Isle of Mull, that hass aal the points o' a Poltolloch terrier, even to the black nose and the cocked lugs, and sits up and barks at you.'

'Oh, criftens!' whimpered Sunny Jim, in agony, dabbing his face incessantly with what looked suspiciously like a dish-cloth; 'I've see'd midges afore this, but they never had spurs on their feet before. Yah-h-h! I wish I was back in Gleska! They can say what they like aboot the Clyde, but anywhere above Bowlin' I'll guarantee ye'll no be eaten alive. If they found a midge in Gleska, they would put it in the Kelvingrove Museum.'

[Munro, *Para Handy*]

Regional news has come from a poll of workers on twenty-two forestry estates. 'No effect beyond discomfort' was reported to George Hendry and Gunnar Godwin by the axemen of Loch Tummel and Loch Tay, in contrast to 'Re-scheduled work necessary, operations curtailed or abandoned for several periods in the season' in the wetter, more westerly areas of Kintyre, Lochaber and Wester Ross. Ardgartan, Loch Awe and Fort Augustus fell into the 'rarely serious' category, although conditions were 'often uncomfortable'. More than half of the nationally-owned forests are in the high rainfall west.

The worst forestry tasks for midge attacks are those requiring both hands like draining, planting, weeding, fencing and tree-felling, i.e. most of them: 'The use of helmets and visors also appears to attract midges, possibly from the odour of sweat from poorly ventilated headwear. Although exhaust fumes from chain-saws may deter attacks, at least temporarily, vivid accounts of resumed assaults during re-fueling were provided by several respondents.' All that and the threat of privatisation in a rural industry which increasingly recruits seasonal workers. Forestry workers are least troubled during chemical spraying and ploughing – the latter providing 'a protective cab as well as mobility'. [*Scottish Forestry*, 1988]

If Kintyre is bad for forestry workers, the story of 'a Highland home in Cantyre' offered a warning to artists as well:

> We are exceedingly fortunate in being able to sit here in perfect peace and enjoyment, without being irritated beyond endurance by the bloodthirsty attacks of swarms of midges and gnats. Thus my little illustration of one of the 'pleasures' attendant upon sketching in the Highlands truly depicts a very common occurrence, and one which, while it drives artists to the verge of frenzy, also compels them to adopt mosquito-curtain veils and other extraordinary head-gear, partially protected by which shrouding they may paint under difficulties.
>
> One of Mr Leech's inimitable sketches in 'Punch' will doubtless be called to mind, where two artists in the Highlands are thus represented with their heads wonderfully done up in gnat-defiers, in which are glazed eyelet-holes and a mouth piece, through

> which a sanatory pipe may be smoked.
> [Cuthbert Bede, *Glencreggan*, 1861]

Places affected include golf courses, as in July 2004: 'The Barclay's Scottish Open provided the usual heady mix of brilliant golf, breathtaking scenery and belligerent midges.' (Ian Wood, *The Scotsman*, 12 July 2004). Also mountains: 'The inevitable happened when I was about thirty feet from the belay, traversing a new slab above the overhang we had just surmounted and making for the point where the crag petered out in a series of vertical steps and grassy ledges. As I was struggling to arrange yet another dodgy nut placement the wind died completely and within seconds a series of intense hot needle pricks told me that the midges had found their man.' And theatre: 'The Widow Dido's All Stars have cancelled a performance of *The Tempest* on Rannoch Moor . . . the Scottish midge too powerful for Prospero's magic to overcome.' The Mull Little Theatre (with forty-three seats) sought Lottery funding for a midge-eating machine so that patrons could go outside during intervals.

BE AFEARD;
THE ISLE IS
FULL OF MIDGES

ARGH, QUICK,
BURN
SOMETHING

HEY, YOU
KNOW YOU
COULD
JUST USE
MUGWORT

Remedies

The word 'midge' derives from the Old Norse *muggia* (*mygg* to the Swedes) so it is no surprise that the oldest folk remedy is mugwort. It was recommended against insects of all sorts, either as a lotion or burnt in a fire. Over the years many repellants have been tried. If you go down in the woods and stop for a picnic lunch, forest lore says that a cedar tree will provide the best shelter from midges because of the thuja which is found in both red and white species.

A mixture of 1 part oil of lavender and 20 parts elderflower water was said by *Home Gardening* to be 'very efficacious', one application lasting a whole evening. Another remedy is based on 1 oz oil of cassia (extracted from cinnamon bark) with 2 oz camphorated oil. All this is mixed with 3 oz lanoline (the fatty basis of ointments extracted from sheep's wool – though sheep suffer from midges) and 'paraffin wax to stiffen' – the upper lip?

Some time ago *Canadian Entomologist* prescribed 1 part oil of thyme, 2 parts concentrated extract of pyrethrum flowers in 2 parts mineral oil and 5 parts castor oil. Oil of white birch (easy to obtain at the lake while making birch-bark canoes) was quite effective. Australians rely on a can of beer. Once the contents have been disposed of in the normal way it is half filled with vegetable oil, together with a quarter cup of antiseptic and topped up with water – salt water, since this recipe comes from the deep sea fishermen of Queensland. Application is typically Aussie: shake the can, splash it on – no worries.

Lemon juice has always had supporters, particularly in the combination 1 oz citronella with 4 oz petrol, 'as least injurious to the skin', but it turns forestry workers yellow. Green is the environmental colour. Mosi-guard Natural is marketed as an alternative to repellents based on DEET (diethyltoluamide) which is dangerous in the bloodstream, especially to children. The natural version of Mosi-guard, successfully trialled against African mosquitoes, is mainly eucalyptus oil. Zoology students exposed Mosi-guarded arms alongside unguarded controls – who had to show great self control. There was a rota among the three of them. Up to 187 bites were recorded per exposed arm (in ten minutes) compared with six on the treated arm. The students used plastic tubing to suck insects off the skin. They then blew midges into a jar of alcohol for counting.

The consumer magazine *Which* has applied its 'more blobs the better' to 400 brands of insect repellent, on sale as Lotions, Sprays, Sticks and Wipes. Sprays are easy to use – Mosi-guard and Z-Stop the winners here. Autan and Wild-Life are the best stick repellants, but no wipe received the *Which* seal of approval since none lasts as long as the average early evening midge attack – enough to dash from car to B & B, though. Smoking insect coils get some support, ultraviolet lamps hardly any. Greatest scorn is reserved for buzzing gadgets meant to mimic amorous male midges: 'Don't buy them – they do not work.'

Journalist Ruth Wishart went to an unlikely factory on the shores of Loch Long to meet the man who invented Shoo: 'Paton Cumming's shed has clearly seen better days from a strictly aesthetic point of view. But it has rarely been busier.' His uncle was involved in developing what became DEET for short-trousered soldiers in North Africa but 'ICI this is not. Inside, Alan and Jessie are packing ever-growing supplies of Shoo. Christened by Paton's grand-daughter, it has found favour with the Forestry Commission, British Airways, British Telecom, and the lesser spotted gardeners of West and Central Scotland.'

Several new remedies have come under discussion, from squashed tomatoes (applied externally) to the oil of the Indian Neem tree which is backed by the Blackwell research team. *The West Highland Free Press* recommends a spoonful of Marmite a day spread thinly on toast – something new for B&B mornings. Vitamin C tablets are also favoured. Perhaps the most surprising form of protection, unadvertised except by word of mouth, was Avon Skin So Soft.

REMEDIES

The Hebridean John Macleod emphasised the importance of keeping two old remedies apart: 'Highland tinkers used to swear by neckties soaked in paraffin – 100 per cent effective in most civilized society. I myself used to rely on the fumes of Capstan Full Strength cigarettes, which did for midges what the doctors said they did for your lungs. I stayed a safe distance from tinkers, and their scarves, while indulging.' *Vital Spark* again:

> 'I promised I would go up and see Macrae the nicht, said Macphail. 'But it's no' safe to go up on that quay. This is yin o' the times I wish I was a smoker; that tobacco o' yours, Dougie, would shairly fricht awa' the midges.'
>
> 'Not wan bit of it!' said Dougie peevishly, rubbing the back of his neck, on which his tormentors were thickly clustered. 'I'm beginning to think mysel' they're partial to tobacco; it maybe stimulates the appetite. My! aren't they the brutes! Look at them on Jim!' With a howl of anguish Sunny Jim dashed down the fo'c'stle hatch, the back of his coat pulled over his ears. 'Is there naethin' at a' a chap could dae to his face to keep them aff?' asked the

engineer, still solicitous about his promised visit to Macrae.

'Some people'll be sayin' paraffine-oil iss a good thing,' suggested the Captain. But that's only for Ro'sa' mudges; I'm thinkin' the Arrochar mudges would maybe consider paraffine a trate. And I've heard o' others tryin' whusky – I mean rubbed on ootside. I never had enough to experiment wi't mysel'.'

Stirling is the gateway to the Highlands and the obvious place for an annual Midgie Festival Competition: 'Retell your worst experiences with the pernicious Scottish midgy and other man-eating insects in cartoons or humorous tales or verse.' The organiser Eric Allison receives a bulging postbag every year from tourists who have banged the gate behind them, relieved to be back in the Lowlands. One tale, perfectly true, begins: 'Why, you might ask, were my husband and I, with Siamese Cat on a lead, wandering around the streets of a Scottish town in our pyjamas at four o'clock in the morning?' Best poem? 'They flew oot o' Hell's open windae!' just shades it from 'Let us Spray'.

Some people are luckier than others, but biochemistry can help:

> Since we made our home near Inverness twelve years ago my husband was very seldom bitten but I had only to step outside the door to be attacked by voracious swarms. However after several very 'irritating' seasons I read a letter in a Sunday paper in which the writer said he couldn't understand what all the fuss was about. He took a strong Vitamin C tablet – 200 or 250 mg – before going out and he never got bitten by midgies. I tried this and found that, for me, it works splendidly.

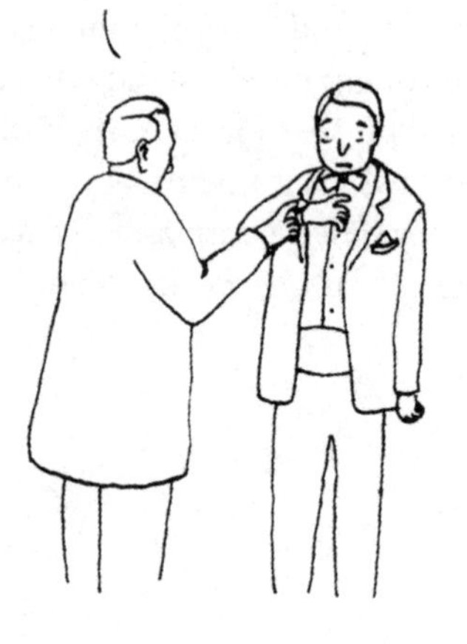
SIMPLY PRESS THIS BUTTON AND A SHIELD OF AVON SKIN-SO-SOFT WILL BE ACTIVATED

Finally

Of all the insects that have raised bumps and itchy spots on me and drawn blood from my cringing skin, I suppose the Highland midge has caused me the greatest melancholy. I can remember standing on a windless Celtic summer evening watching the hidden sun ring black Mull in a lake of fire. It was a time of prose inspiration and poetic derivation, lofty thoughts about man's place in the universe, and a hush like a benison, all ruined by the fact that local midges were working overtime and making me feel that hundreds of tiny hypodermic needles were being plunged into my face.

Like most people, I have tried a witch's brew of creams and other repellents to make midges, if not curl up and die, at least pause in their tracks. All to no avail. Ancients, wise in the ways of midges, told me that smoke from a pipe tobacco called Navy Cut plug was effective in thwarting them temporarily – rather like a destroyer making smoke for protective purposes – but that fag smokers just suffered. That was not strictly true. Once I smoked black Burmese cheroots and had crafty drags at Balkan cigarettes, both of which gave out smoke redolent of Eastern bazaars, seraglio nights and a faint whiff of lodging house cat. The midges seemed stunned by the effects of passive smoking and one sensed that they were suddenly coughing and wiping their eyes. After a pause of about 20 minutes, though, they seemed to be getting used to the scent, perhaps even inhaling it with enjoyment, and then having a quiet drink off you as a chaser. (Albert Morris, *The Scotsman*)

An early midge-eating machine was reported in *The Times* by Ben Macintyre in 2002:

> Last year, shortly before leaving the US and returning to Britain, I happened to come across an advertisement in an in-flight magazine for a revolutionary new machine called a Mosquito Magnet . . . The technology seemed fairly simple: a green box with a sort of trumpet contraption attached. The machine is powered by propane and emits a thin, moist plume of carbon dioxide intended to mimic the exhalation of a hot-blooded

> animal. This is flavoured with Octenol, a chemical compound barely perceptible to man but foie gras to a midge. Octenol, the company says, 'is a naturally occurring by-product that comes from some animals (pheromone and kairomone) such as oxen and cows, which ingest large amounts of vegetable matter.' Think: distilled cow burp. Lured by the illusion that there is a vast, heavy-breathing, hot-blooded cow in the vicinity, the female midges swarm towards the source, where they are sucked by vacuum into a mesh bag and then perish . . . Sometimes at night in Scotland, when the wind is still, I can hear the faint purring of my midge machine as it steadily devours these beasties which, for too long, have devoured me. And I am content.

This American product had free publicity by being used at Loch Laggan for the TV series *Monarch of the Glen*. According to midge-terminator.co.uk, 'The effect is significant after 4–8 weeks of continuous use. Happy customers in Scotland have reported regularly trapping over one million midges in a night and 2,500,000 midges (21/2 lbs in weight) were collected in one net! . . . Over 1,000,000 Mosquito Magnet machines have now been sold worldwide in the last six years! 17,000 machines are currently manufactured daily seven days a week. The manufacturer, American BioPhysics Corp., has been named Number One on "2003 Inc. List of the Fastest-Growing Private Companies" in the United States. By comparison Midgeater UK by Calor Gas and the Dundee engineering firm Texol sounds couthily Scottish. A spokesman of the Scottish Tourist Forum said that 'if it does all it says it will make life bearable for tourists in beer gardens and luxury hotels'. Ninety-six five-star time-share properties were opened on the banks of Loch Lomond. This followed the identification of breeding areas on the 300-acre estate. The machines, emitting 'carbon dioxide, octenol and a thermal lure', went under the name Dragonfly Professional. One and a half thousand Midgeaters were quickly sold.

However, progress towards a midge-free Scotland came under the threat of legal proceedings. BioPhysics claimed that Midgeater had breached their Mosquito Magnet patent, and the Scottish manufacturer admitted stripping down the American model for inspection. Propane-powered carbon dioxide and octenol are common to both, although clearly Dr Blackwell's team came up with this 'midge bait' out of their own

AARGH

research. But there has been press confusion in Scotland over 'electric zappers', when it is clear that Calor-generated electricity is linked to the Midgeater's vacuuming function. Nowadays no summer passes without something new on the midge front. Hot dry ones have reduced numbers without need of man's intervention. But invention is not lacking. The simplest (if short-range) solution is adapting the waft of hot air which greets customers at the supermarket. A 2009 headline suggests a use for dead pests: 'Revenge is tweet. Midges are recyled into bird food.'

Scientists continue active. Professor Jenny Mordue of Aberdeen University and her team set up their own midge traps while doing new work on repellents. Prompted by the fact that some people do not attract midges, attempts have been made to make their deterrent emanations available to others. Ketones are used in the latest brand-leader Smidge, and the Aberdeen approach includes a 'puffer' which sends out natural chemicals ahead, for example, of running athletes. According to Dr Mordue, 'This is something entirely new. The device stops midges coming towards you, whereas with all the other repellents on the market the midge has to land on the skin before they work.' Also new on the market are midge patches from which the body absorbs thiamine or Vitamin B1. The resulting aroma is pleasant to humans but turns midges away.

My earliest ideas on midges took no account of climate change, until the cold 17th century came to mind. It now seems from the ever-developing science, however, that cold winters do more harm to bat predators than to larvae in frozen ground. Warmer summers have certainly made it possible for up to three hatches during the season. Global warming might extend that into the autumn, but hot dry summers would not suit these insects. Perhaps the mosquito population would rise. Weather changes in the Horn of Africa arising from heavy rain and cyclones in 2019 caused beyond-biblical plagues of locusts.

Continuing this return to the PERSONAL opening section, I also missed Bonnie Prince Charlie's biting experience – under chilling rain – in Benbecula before the much sung-about crossing to Skye. He had been helped through bogs so deep that a companion had to search to his shoulder for one shoe after another. Journal-keeping visitors were never so fully immersed in the environment. As Tim Kirby's cartoons make clear, this little book is intended to amuse but the subject is serious. Journalists, summer by summer, mix humour with whatever they can glean from scientists. Neither group thinks about midges in relation to what were historically immense changes in land use. Jim Gilchrist of *The Scotsman* – uniquely – picked up on some of this and brought in the removal of families from clachans to make way for sheep. Beyond that, many abandoned cultivation ridges recall the hard work of drainage once carried out in a heavily populated Highland region.